INTERNATIONAL SPACE STATION

1998–2011 (all stages)

Contents

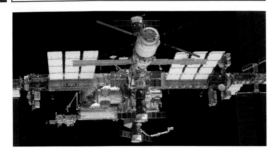

OPPOSITE Set against an ice-clad portion of the Earth's surface, veiled only sightly by clouds more than 200 miles below, the completed International Space Station is a majestic tribute to the work of 15 countries over several decades. *(NASA)*

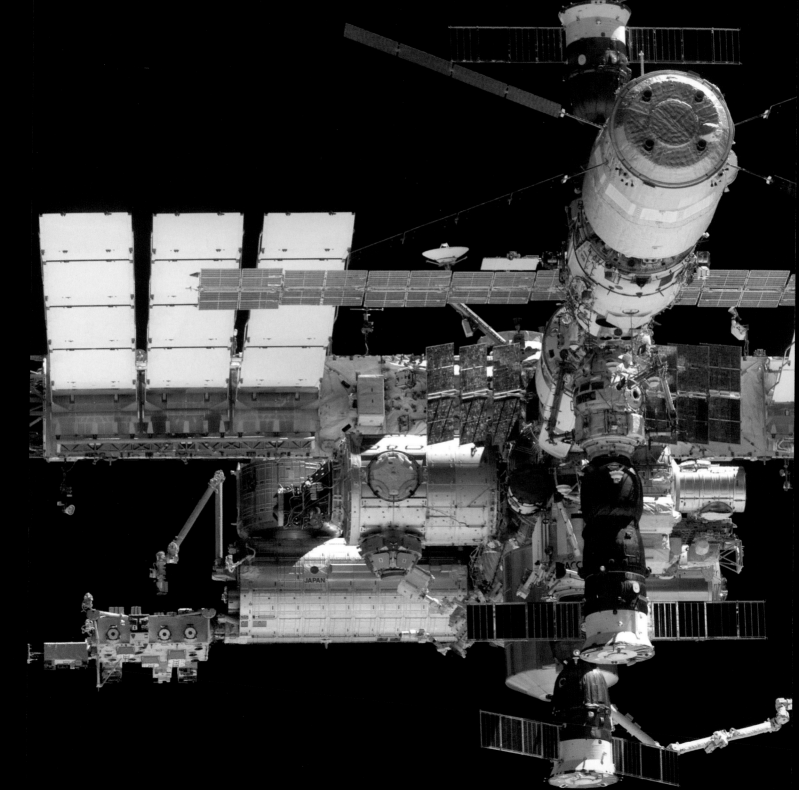

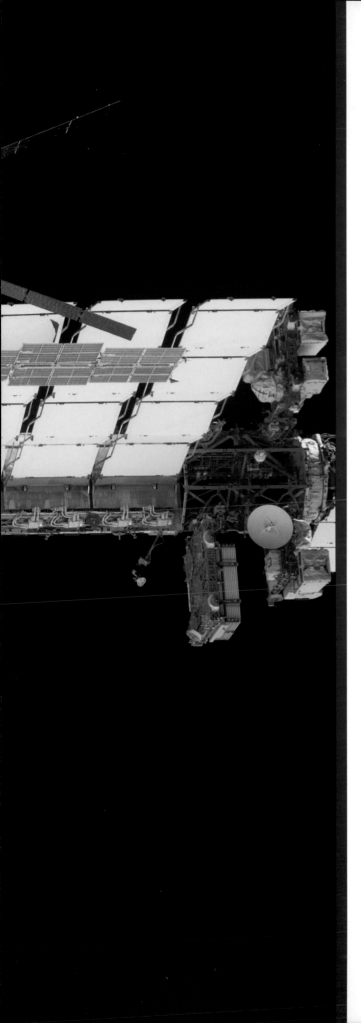

Introduction

The International Space Station (ISS) has been long in coming and has taken more than 12 years to build, but it has united 15 countries and five space agencies in an unprecedented effort to construct a giant orbiting facility that may outlast many who helped design it.

OPPOSITE With Russian Soyuz manned taxi spacecraft and Progress cargo-tankers top and bottom and Europe's Automated Transfer Vehicle in the foreground, the International Space Station is a truly global science platform. *(ESA)*

ABOVE **The Shuttle STS-126 crew (red T-shirts) poses with the six-member ISS crew.** *(NASA)*

BELOW **Canada's robotic arm projects from the ISS over a sun-drenched planet below.** *(CSA)*

Forged from the tools of a Cold War, the ISS is a testament to cooperation, not only between amicable partners but also between former adversaries, bridging the ideological divide that once seemed permanent. Historians of both politics and technology assert that herein is its true value. Where once it was said that after landing on the Moon, for mankind only the limitations of imagination acts as a constraint on the possible, now it can be said that if the world can unite behind this great international venture, surely a world divided by differences must become a thing of the past.

Yet the ISS is more than a complex assembly of modules, truss structures and solar wings. It is the culmination of efforts in many countries for humans to remain in space for long periods, experience that dates back to the early 1970s when both Russia and the USA launched the first habitats capable of supporting people for long periods in the challenging environment of weightlessness. The ISS is a direct product of those early attempts to build a human presence in space and as such it builds on early efforts without which the success of the International Space Station would be impossible. The Russians were first with their Salyut space station, launched in 1971, followed by the Americans with the Skylab station just two years later. A major participant in robotics on the ISS, Canada contributed first by building the Shuttle's remote manipulator arm, first flown in 1981 on the second Shuttle flight. And European expertise was built upon participation when Germany led a consortium of nations to build the first orbital laboratory called Spacelab, first carried by the Shuttle in 1983.

Today, the ISS is complete in its definitive design configuration but it will continue to change and adapt to the needs of scientists and research workers around the globe who routinely visit the station, use its tools and profit from the results of myriads of experiments inside and outside this 400-ton (363,000kg) complex. It is assured a further life until at least 2020 but participating agencies are already talking about maintaining the station until 2028 or later, when new generations of spacecraft may use it as a springboard to the deeper regions of the solar system. From this facility research will benefit people on Earth and provide information for engineers planning mankind's first forays beyond the Earth–Moon system, perhaps as far as Mars and the asteroids. New processes for medicines, materials and biological agents are even now being tested and from the ISS we may learn more about the human body and its ageing process than we ever could by remaining on Earth.

Uniting nations in a cooperative venture where interdependence replaces isolation, where convergence replaces conflict, and where future aspirations are shared and not opposed, allows a new generation of young idealists to work in a truly global village to suppress the petty hatreds of bitterness and rivalry that once fuelled the nascent space programme, where competition

AN IDEAL HOME IN SPACE

In one of the more unlikely steps to building a space station, an exposition in London, England, known as the Ideal Home Exhibition, would be the venue for the first public display of the way NASA would get its first station in space. With space on the world agenda and public interest at a high, the UK's *Daily Mail* newspaper sent a reporter to visit US aerospace companies and find out what they thought about future possibilities of living in space and what such stations would look like. He visited the Douglas Aircraft Company in Santa Monica, California, who were only too pleased to publicise their work on such an application.

Several noted rocket engineers, including von Braun, had proposed the use of spent rocket stages as an economical way of putting a station into space. Retain the last stage of a Saturn rocket, vent the residue of its remaining fuel into the vacuum of space and convert the cavernous interior of the propellant tank into a habitable place to work. Pressurised with oxygen and nitrogen, it could be made to support teams of astronauts ferried back and forth by a multi-man successor to Mercury or a space-plane like the Air Force Dyna-Soar.

Douglas liked that idea and, inspired by

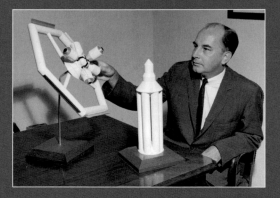

LEFT NASA's Rene Berglund in 1963 with a model of a space station concept comprising linked modules launched by a giant rocket (foreground). *(NASA)*

the nudge from one of the UK's leading newspapers, a team was set to work under W. Nissam with a budget of $10,000. The company already had the contract to build the S-IV second stage for Saturn and in their concept adapted that as a potential habitat in space. Four petal-like doors covering the forward section which supported the manned spacecraft for launch would open in orbit, revealing solar cells for electrical power.

The newspaper liked the idea and asked Douglas to build a full-scale replica of the concept for public display at the *Daily Mail* Ideal Home Exhibition in March 1960, a full-size structure through which members of the public could walk – and 200,000 did just that, getting their first experience of what a space station might look like!

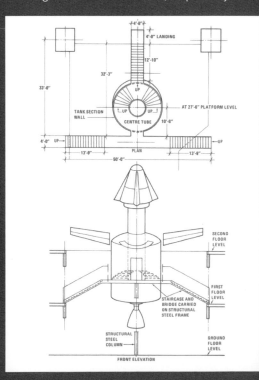

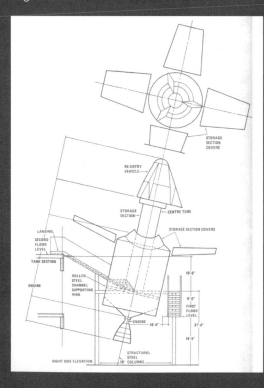

FAR LEFT The *Daily Mail* inspired a space station concept put together by the Douglas Aircraft Company for the London Ideal Home Exhibition. *(David Baker)*

LEFT The Douglas space station concept was set up as a full-scale, walk-through, replica. *(David Baker)*

is a proud thing and not a springboard to violence. As cargo ships rise from launch pads in Japan and from the coast of South America, as freighters lift off from the sprawling complex at Baikonur and as commercial companies send logistical supplies to the ISS from Cape Canaveral, those who foresaw an orbiting space station as an opportunity for mankind to work together in peace and harmony will be pleased with what they started.

Stepping stones to a habitat

The tortuous path that NASA (National Aeronautics and Space Administration) would follow to build a space station in orbit was laid in politically motivated decisions set by presidents and lawmakers. Not for 40 years after it was formed would NASA make a start on its prime goal – constructing a permanent workplace in space. When it came it was very different from the original concept developed from ideas evolved over the preceding century and it would only happen at all because of the collapse of the Soviet Union.

From a Power Tower to Freedom

Tasked with sending men to the Moon by the end of the 1960s, NASA grew far beyond the expectations of its founders. It planned ambitious ventures to far-flung

BELOW Boeing's Space Operations Center envisaged modules for research and satellite servicing. *(Boeing)*

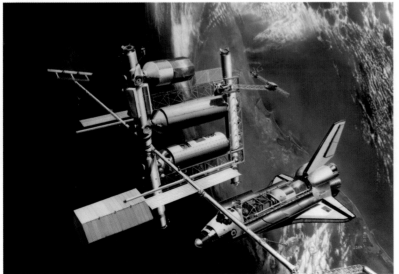

destinations but stuck to an age-old concept of the space station as a springboard to the planets. But in achieving the Moon goal, NASA saw those dreams collapse and searched in vain for the way to build a habitat in space.

On 25 January 1984, at the annual State of the Union address to a joint session of Congress, President Reagan authorised NASA to begin work on a space station 'and to do it within a decade'. In February, NASA administrator Jim Beggs set out to tour the pro-Western world and conduct presentations to invite interest beginning with a visit to the UK, where he was given a cool reception by the Thatcher government. From there he visited West Germany, Italy, and France before flying to Japan. Continental European countries gave Beggs a warm reception and West Germany in particular was keen to participate in the space station as part of the European Space Agency (ESA). Japan too was especially excited about the prospect of making further progress in its national space effort and, having already designed and produced the Shuttle's robotic arm, Canada was eager to develop advanced robotics for the station.

The configuration incorporated three elements: a central core structure of four laboratory and habitation modules supporting a crew of eight in a 28.5-degree orbit; an unmanned co-orbiting platform for free-flyer experiments; and a second unmanned free-flyer orbiting the Earth at 90 degrees to the equator. But as Jim Beggs gathered his list of partners, the design was already in a state of flux. Throughout the summer three design options were considered: the 'Power Tower', the 'Planar' station and an odd configuration known as the 'Delta' station. Originated by McDonnell Douglas and Grumman, the Power Tower consisted of a 300ft (91m) tall lattice tower across which would be attached a 200ft (61m) wide cross-beam mounted two-thirds of the way up, with four Solar Array Wings at each end. Up to five modules, each up to 35ft (10.7m) long, could be attached to the bottom of the tower with large radiator wing panels on either side. This configuration would always fly with the modules pointing Earthwards and achieve some stability from a pendulous alignment with the Earth's gravity. The Planar station concept consisted of a single rigid truss assembly 300ft (91m) long with the modules attached at the centre and four large Solar Array Wings at each end. Science experiments would be attached to a large A-frame extending 80ft (24m) above the truss. The Delta station derived its name from an inverted triangular shape at the top of which would be a 175ft by 125ft (53m by 38m) platform of solar cells covering 28,600ft² (2,657m²). Extending down from this platform would be an inverted V-shaped field of beams supporting five pressurised modules at the bottom.

Within months the Power Tower became the reference baseline. Europe proposed a pressurised experiment module it would call 'Columbus', lifted to the station by Shuttle where it would form one of the cluster of modules, including those from the US and one from Japan. By mid-1985 the station configuration had shifted from the Power Tower concept to a dual-keel design, a configuration increasingly favoured since the beginning of the year. This envisaged a transverse truss holding solar arrays midway across a rectangular lattice structure. The total span of the solar array truss would now be 503ft (153m), supporting at its centre the pressurised living and work modules.

There would be only two US modules, each increased in length from 35ft to 44.5ft (10.7m to 13.4m) but retaining a diameter of 13.6ft (4.15m), attached in proximity to the modules from Europe and Japan. The parallel long sides of the rectangle supported by the truss would each be about 330ft (101m) tall, 126ft (38m) apart and connected at the bottom and at the top. Overall height would be 361ft (110m). The top cross-section would overhang the vertical sides to provide a total span of 297ft (91m), providing space for instruments and equipment. The upper cross-section would be used for astrophysical observations and the lower one for Earth observations with the idea being to 'fly' the station perpendicular to the path of the orbit.

The dual-keel design would have only four solar array wings, adding dish-shaped solar heat collectors to drive alternators producing 400-cycle AC electric power. This reduced the large surface area of the station, causing drag from the random air molecules present even in low Earth

RIGHT Large payloads could be placed around the rectangular-shaped truss structure for ease of access by the Shuttle. *(NASA)*

LEFT In 1984 NASA got formal approval to begin development of a space station, this concept known as the Power Tower put all the solar arrays at one end of the assembly. *(NASA)*

BELOW By 1986 the design had shifted to the Dual Keel concept but retained provision for satellite servicing. *(NASA)*

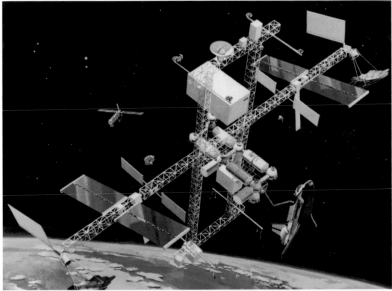

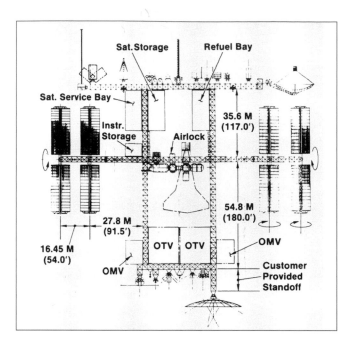

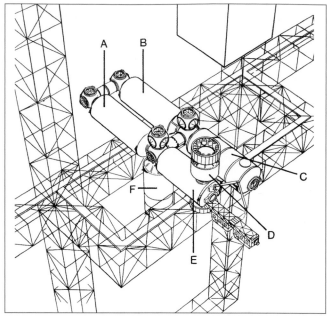

ABOVE The Dual Keel design had provision for additional space 'tugs' (Orbital Transfer Vehicle and Orbital Manoeuvring Vehicle) to fetch and return to distant orbits payloads that could be worked upon at the station. *(NASA)*

ABOVE RIGHT The Dual Keel configuration was ambitious and ultimately seen as unaffordable. Key: A) US Habitation Module; B) US Microgravity Lab; C) Europe's Columbus module; D) Japanese Logistics Module: E) Japanese Experiment Module; F) US Logistics Module. *(NASA)*

RIGHT By 1989 the station was in crisis, A redesign took away everything but the central truss to which modules could be attached at its centre. *(NASA)*

orbit. The problem with the Power Tower had been instability. This was avoided by the dual-keel design, which, with the modules located at the centre, had better balance. Europe got to work designing its Columbus module while Japan began a detailed design of its own module.

When approved by Reagan in 1984 it had been hoped to build the station with 8–10 flights, but that estimate had now increased to at least 19, and possibly as many as 31 flights. Questions also emerged regarding the amount of space-walking, or EVA (extra-vehicular activity), necessary to build and maintain the dual-keel station and over the need for a lifeboat rescue vehicle. As 1989 drew to a close it looked as though the station, named 'Freedom', may be on track for a first element launch in March 1995. But then a special review group concluded there was an impossibly high reliance on EVA, which is inherently dangerous and saps valuable work time otherwise spent on science experiments. Just to keep the station going, said the group, astronauts would have to expend 2,284 hours each year, an average of

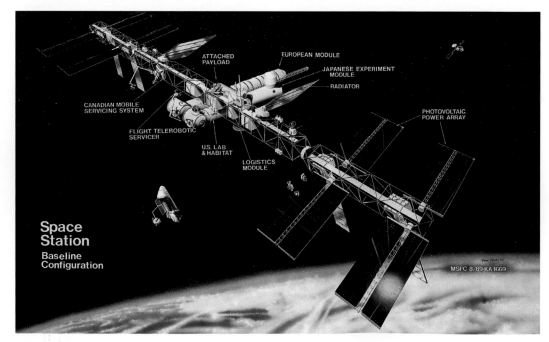

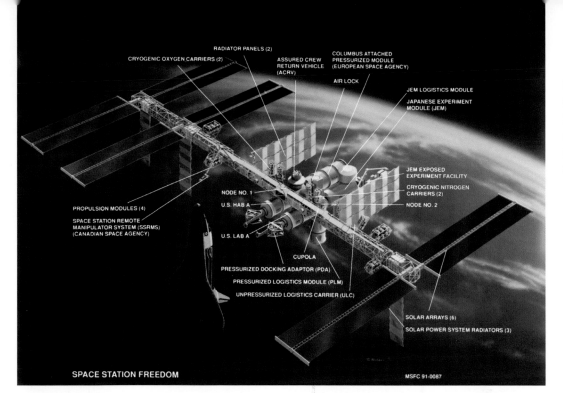

RADIATOR PANELS (2)
CRYOGENIC OXYGEN CARRIERS (2)
ASSURED CREW RETURN VEHICLE (ACRV)
COLUMBUS ATTACHED PRESSURIZED MODULE (EUROPEAN SPACE AGENCY)
AIR LOCK
JEM LOGISTICS MODULE
JAPANESE EXPERIMENT MODULE (JEM)
JEM EXPOSED EXPERIMENT FACILITY
CRYOGENIC NITROGEN CARRIERS (2)
NODE NO. 2
NODE NO. 1
U.S. HAB A
PROPULSION MODULES (4)
SPACE STATION REMOTE MANIPULATOR SYSTEM (SSRMS) (CANADIAN SPACE AGENCY)
U.S. LAB A
CUPOLA
PRESSURIZED DOCKING ADAPTOR (PDA)
PRESSURIZED LOGISTICS MODULE (PLM)
UNPRESSURIZED LOGISTICS CARRIER (ULC)
SOLAR ARRAYS (6)
SOLAR POWER SYSTEM RADIATORS (3)
SPACE STATION FREEDOM
MSFC 91-0087

LEFT By 1991 Space Station Freedom had settled into the configuration concept it would eventually become. *(NASA)*

one two-person EVA every other day. By July the estimates for maintenance alone exceeded 3,276 hours per year – five two-person EVAs per week – an impossibly unacceptable target. The Johnson Space Center set about a rigid series of changes to drive annual EVA maintenance down to just 507 hours.

On 21 March 1991, NASA sent its new plan to Congress. Overall station span would be reduced from 493ft (150m) to 353ft (108m) and each of the two US modules would be downsized in length from 44ft (13.4m) to 27ft (8.2m). Whereas the modules were to have been fitted out in space, the smaller size would allow them to be assembled and launched as complete elements. They would be connected by nodes, barrel-shaped multiple docking ports serving as passageways and outfitted with equipment essential to running the station. An airlock module would also be attached to one of the node ports to allow astronauts to conduct EVA without depressurising the entire station. Instead of constructing the long truss assembly in orbit from a bundle of parts carried to orbit by the Shuttle, truss sections would now be preassembled and fitted out with conduits and cable trays on the ground, for launch as completed elements. This alone would cut EVA time by 50 per cent, requiring the crew to hook up the connections rather than erect the structure in space.

The number of Shuttle flights necessary to build Freedom had been cut from 34 to 17

but the date of the first element launch had been put back by one year to March 1996. Man-tended capability would be reached in December 1996, with Freedom complete and permanently manned with a reduced crew of four by late 1999. Already, some 79,000 workers were employed by US companies to build the various elements across 39 US states, added to which were aerospace workers in Europe, Canada, and Japan. But when Bill Clinton entered the White House in January 1993 he ordered NASA to redesign the station yet again! Over the next several months a blue-ribbon review committee set up on 10 March under former Apollo executive Dr Joseph Shea examined a wide range of configuration designs that had not been looked at before – designs that substantially downgraded the station and its capabilities to fit within budget limits.

On 17 June the president made a public statement backing a low-cost configuration known as 'Space Station Alpha' and on 24 June Clinton ordered NASA to develop a new programme plan. But on 2 September Al Gore, the vice president, sat down with the prime minister of Russia, Viktor Chernomyrdin, and signed an agreement embarking on a joint effort to design and build a completely new International Space Station. The origin of this transformation goes back to 1986, just two years after the decision by Ronald Reagan to give NASA the go-ahead to build a station.

Chapter One

A permanent place in space

While American astronauts went to the Moon and US engineers built a reusable Shuttle, Russia's cosmonauts were learning how to live and work in space. Since the early 1970s the Soviet space programme had assembled an impressive record in long-duration flight – something NASA was desperate to emulate.

OPPOSITE Mir grew over its 15-year life with additional modules and equipment, providing a base on which both US and Russian crewmembers learned how to work in space together. *(NASA)*

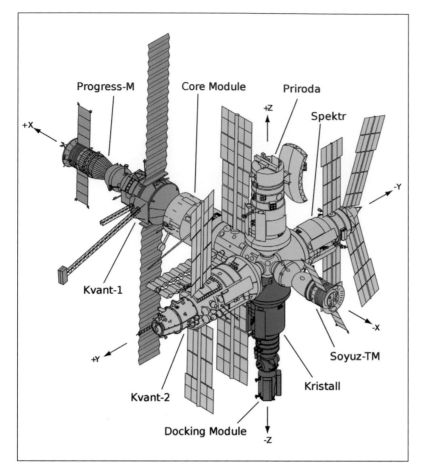

Progress-M · Core Module · Priroda · Spektr

+Z · +X · -Y · -X · +Y · -Z

Kvant-1 · Soyuz-TM · Kvant-2 · Kristall · Docking Module

ABOVE Mir provided the engineering experience for Russia's contribution to the ISS and much was learned by NASA from the experiences of Russian engineers and cosmonauts. *(NASA)*

BELOW Salyut 7 and Cosmos 1686 provided valuable experience in docking very large structures in space. *(NASA)*

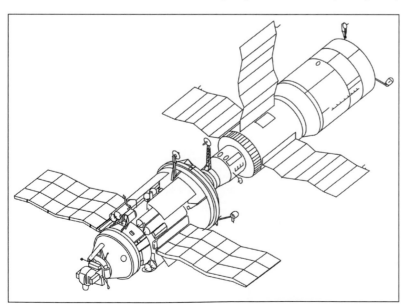

Mir

The launch of the Soviet space station Mir ('peace') came on 19 February 1986, just over three weeks after the loss of *Challenger* in a disaster that kept the Shuttle grounded until 29 September 1988. The Mir core module had been developed out of the Salyut series of space stations first launched in 1971, developed as a long-duration facility launched by a Proton rocket. Mir presaged a new capability for keeping people in space, using the building-block approach to expand the facility into a fully equipped laboratory.

To achieve that, while looking similar to the second generation stations Salyut 6 and 7 first launched in 1977, Mir had a forward docking section incorporating five ports – one at the extreme forward end and four radial ports at 90-degree intervals around the circumference. This docking section incorporated a transfer system called Lyappa, which would latch on to a new module docked at the extreme forward port and rotate it around 90 degrees, reattaching it to one of the four radial ports, thus freeing the forward port for another module.

Up to four permanent plug-in modules could be delivered to the forward node, each moved round to a radial docking ring. And just like Salyut, there was another docking port at the rear of the Mir station making six in all in a system that could accommodate four new large experiment modules while retaining respective docking ports for manned Soyuz and unmanned Progress cargo-tankers.

Internally, Mir had a single pressurised work compartment of 1,400ft³ (40m³), a length of 25ft 2in (7.7m), and a diameter of 9.5ft (2.9m) in the forward section and 13.75ft (4.19m) in the main body of the compartment. The rear of Mir comprised a service propulsion section with the same diameter as the aft work compartment and a length of 7.5ft (2.29m). It supported two 660lb (300kg) thrust rocket motors for re-boosting altitude and making minor orbit changes. Access into Mir from this end was made via a central tunnel 6.5ft (1.98m) in diameter connecting the aft docking port to the main work compartment.

Total length of the core module as launched was 43ft (13.1m) with a mass of 46,000lb

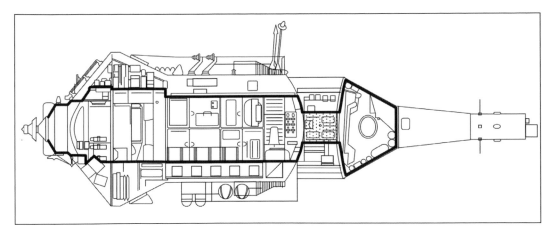

LEFT Designated Cosmos 1686, this TKS module formed the basis for separately launched modules attached to Salyut 7 and later the Mir space station. *(NASA)*

(20,865kg) and attitude control handled by 32 control thrusters. About 9kW of electrical power came from two solar array panels with a total span of 97.5ft (30m) covering 820ft^2 (76m^2), 26 per cent more than arrays on Salyut 7. Added to which, each new module would carry its own solar cell arrays to supplement power for the science equipment they contained.

The first visitors to Mir were cosmonauts Leonid Kizim and Vladimir Solovyov, launched on 13 March 1986, followed six days later by an unmanned Progress cargo-tanker which remained docked at the station until 20 April when it was de-orbited, the first of four Progress cargo-tankers launched to Mir over the next 12 months. After 50 days aboard Mir, in early May Kizim and Solovyov used their Soyuz T-15 spacecraft to fly across to the Salyut 7 space station about 1,700 miles (2,736km) away in an adjacent orbit, and became the first crew to visit two space stations on the same mission. For some 50 days more they carried out tests and retrieved some equipment before returning to Mir on 27 June. After four months in space Kizim and Solovyov returned to Earth on 16 July 1986, leaving Mir unmanned for more than six months.

Manned operations resumed on 5 February 1987, with the launch of Romanenko and Laveykin aboard Soyuz TM-2 followed by the first plug-in module, Kvant-1 (Quantum-1) on 31 March. Kvant-1 was designed as an astrophysical laboratory with various telescopes developed in cooperation with the UK, the Netherlands, West Germany and the European Space Agency. It had a length of 17.4ft (5.26m), a diameter of 14.3ft (4.34m) and weighed about 45,400lb (20,600kg) at launch. After some

difficulty getting a proper connection, Kvant-1 docked to the aft port on 9 April using an automated approach system where it would remain for the life of the Mir station, providing an additional internal volume of 620ft^3 (17.56m^3). A 20ft (6.1m) long propulsion unit, located on the extreme aft end of Kvant-1 and used only for the rendezvous manoeuvres, was separated, reducing the weight of the module to 21,000lb (9,525kg) and exposing a standard docking unit as the new aft port for future Progress and Soyuz vehicles.

For more than two years the station was manned continuously by five more Soyuz crews in succession, including cosmonauts from Syria, Afghanistan, and France, until 26 April 1989, when it was once again left unmanned. In those six expeditions, two-man cosmonaut teams spent an average of four to six months in space before returning a few days after they

BELOW The TKS spacecraft incorporated a recoverable capsule originally designed to carry a crew of three. *(NASA)*

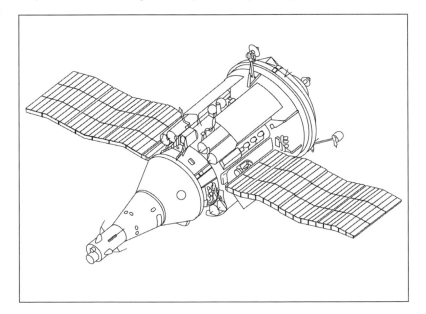

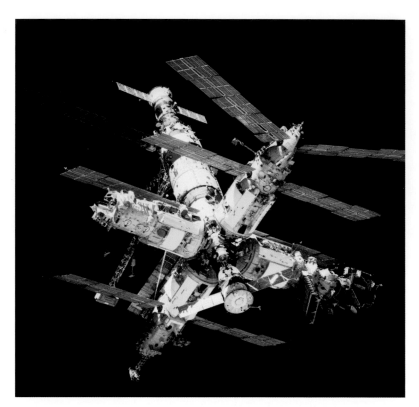

ABOVE The TKS
design origin of the
four separate modules
is evident, arranged
radially around the
central core of Mir.
(NASA)

space flight were conducted at the end of 1991 and on 17 June 1992, President George Bush of the USA and President Boris Yeltsin of Russia signed an agreement pledging cooperation through Shuttle missions to Mir, the initial protocol being signed by NASA and the Russian Space Agency, the RSA, on 5 October 1992. This included the flight of a Russian cosmonaut aboard the Shuttle in 1994, a US astronaut flying on a Soyuz to spend more than 90 days aboard Mir, Russian cosmonauts being changed out via the Shuttle and development of a common (androgynous) docking module for use by either spacecraft. Throughout 1992, NASA and its industry partners began to think of how they could exploit the relaxation of political tension with Russia and utilise their considerable stock and hard-won experience in cutting the costs of the US Freedom station.

The International Space Station

The Russians had been launching stations for 20 years, routinely keeping people in space for between six and eleven months. In the end it was a presidential decision, Bill Clinton wishing to use NASA as a winch to haul in Russian expertise and forge a new working relationship. When the final agreement was signed on 2 September 1993, it probably saved the NASA station and certainly gave it new purpose. The challenge now was staggering in its implication. With a five-team partnership comprising the USA, Russia, Europe, Japan, and Canada, technical integration of separate modules and systems would be a major challenge. There were new languages, fresh working practices and very different engineering approaches, as Russia leapfrogged the junior partners and got straight down to business with the Americans. The names Freedom and Alpha One were defunct; now it was truly the International Space Station – the ISS.

On 7 December 1993, all partners formally secured the agreement for Russia to be a part of the ISS. Russian elements were added to the existing Alpha One concept and modified Soyuz would double as lifeboats, with Progress cargo-tankers supporting what was now an enlarged operational plan for sustained use. Perhaps the

were relieved by the next team. During that period too, crews offloaded equipment, fuel and water from 15 Progress vehicles that had come and gone at the aft port. Meanwhile, Salyut 7 would remain derelict, eventually burning up in the atmosphere in February 1991 and scattering debris over Argentina and Chile. A resumption of flights to the Mir station began with the flight of Aleksandr Viktorenko and Aleksandr Serebrov on 5 September 1989, little more than four months after the first crew returned to Earth, beginning a series of flights that would have the station permanently manned until 27 August 1999.

In that time, the world changed a great deal and what had begun as a race for technological and ideological supremacy between global superpowers would end with Russia free from communism. After the collapse of the Soviet empire Russia lost territory, natural resources, manpower, factories, and government facilities now located in what, almost overnight, became foreign countries. Not least was the massive former Soviet launch facility at Baikonur in Kazakhstan, future use of which was secured under a lease. There was just enough money to maintain the Mir programme but their shuttle *Buran* was abandoned.

Initial discussions to cooperate in human

Chapter Two

ISS Phase One – missions to Mir

Following agreement to merge their efforts at long-duration missions in Earth orbit, NASA and its Russian counterpart set to work planning a series of missions that would prepare the way for assembly of a truly international laboratory in space – one hosting astronauts from several different countries.

OPPOSITE An artist depicts the docking of *Atlantis* to the incomplete Mir in November 1995. *(NASA)*

23

The plan to merge the ambitious human space flight programmes of the United States and Russia was both brave and bold. Brave, because it faced some objections from European countries which, until the decision in 1993 to embrace the newly democratised Russia, had been America's prime partner through its European Space Agency; now Russia was the main partner and the others, including Japan, felt marginalised. And it was bold because the way the Russians went about engineering their space vehicles was completely different to that of the United States; there would have to be much compromise and a lot of information exchange from each side to the other.

In truth each had much to learn from the other and that had been understood for the previous 20 years, when the Apollo-Soyuz Test Project (ASTP) brought astronauts and cosmonauts together for a joint docking flight in a mission launched on 15 July 1975. Just over two days later they docked, visited each others' spacecraft and spent almost two days carrying out joint operations. This flight, accomplished during the Cold War, inspired plans for a mission to link NASA's Shuttle with a Russian Salyut station, but politics got in the way and it never happened.

Now, Phase I of the 1993 agreement would involve Shuttle flights to the latest Russian station, Mir. In that respect it was the next logical step beyond ASTP and the realisation of a plan that advocates on both sides of the political divide had wanted since that first joint endeavour. ASTP had given NASA the opportunity to see for itself how the Russians worked, solved problems and constructed solutions, and each side gained a new and healthy respect for the other's work. Now it was time to put all that to the test. But only by flying together in space could the real test take place.

STS-60

3–11 February 1994

The historic flight of NASA's sixtieth Shuttle mission was a landmark event, being the first US spacecraft to carry a Russian cosmonaut as part of its crew. But it would not rendezvous or dock with Mir. Launched on February 3 1994, STS-60 was commanded by Charles F. Bolden and included in its six-person crew Sergei Krikalev, who participated in this purely scientific flight. During the eight-day mission experiments were conducted, and a commercially developed Spacehab module attached to the forward part of the payload bay allowed several other tasks to be conducted with a special video-link set up between the Shuttle and three cosmonauts aboard Mir.

STS-63

3–11 February 1995

Launched exactly a year after the first US flight carrying a Russian into space, this was the first flight to carry a female pilot – Eileen Collins – and to include the STS-60 back-up cosmonaut Vladimir Titov. The eight-day flight included the first rendezvous with the Mir space station, *Discovery* coming to within 36ft (11m) of the orbiting complex before backing away to a distance of 400ft (122m) to commence a slow fly-around for a photo-survey. A Spacehab module once again provided opportunity for science and technology tests and a set of six tiny sub-satellites with which ground controllers could calibrate radar equipment to more precisely monitor space debris were released. Other tests into how the Shuttle environment affects the space around it were conducted

BELOW Sergei Krikalev (top right) became the first Russian to ride the Shuttle when STS-60 was launched in February 1994. Charles F. Bolden (bottom right) became NASA Administrator in 2009. *(NASA)*

FAR LEFT Mir as it appeared to the crew of STS-63 in 1995 when the Shuttle performed the first rendezvous with the Russian station. (NASA)

LEFT So close and yet so far! From only 36ft, cosmonauts aboard Mir gaze across at the Shuttle. (NASA)

using a SPARTAN platform deployed by the remote manipulator. British-born Michael Foale joined Bernard Harris on a 4hr 38min EVA, Harris becoming the first Afro-American to walk in space.

MIR-18/NASA-1

14 March 1995

Little over a month after *Discovery* returned to Earth, the first US astronaut to ride a Russian spacecraft was launched from Baikonur when Norman Thagard, the first American to visit the Russian station, was carried to Mir along with cosmonauts Vladimir Dezhurov and Gennady Strekalov aboard Soyuz TM-21. They joined Aleksandr Viktorenko and Yelena Kondakova, who had been aboard Mir since October 1994 but would return in Soyuz on 22 March. The unmanned Spektr module launched by Russian rocket on 20 May 1995 carried US instruments and docked to the forward Mir port on 1 June. With a weight of 43,300lb (19,640kg), Spektr was 47.4ft (14.5m) long and had a diameter of 13.45ft (4.1m), carrying remote sensing instruments for Earth and atmospheric studies. Spektr would also provide living and working areas for visiting NASA astronauts. It had two Kvant-2 type solar arrays with a total span of 76.45ft (23.3m) and produced 6.9kW of electrical power. Spektr was adapted

from an abandoned military programme for space surveillance and missile defence and the forward Octava module was replaced with a conical device. Both Spektr and the later Priroda modules were paid for by NASA and equipped with US instruments. Before Spektr docked to Mir the two Russian cosmonauts conducted three spacewalks to complete installation of Kristall's solar wing on Kvant-1 and retract Kristall's second solar wing to make way for *Atlantis* arriving a month later.

BELOW CocaCola extends its global influence to the heavens, as STS-63 astronauts carry light refreshment to the ISS. (NASA)

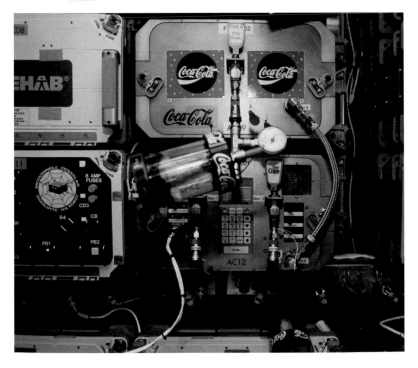

RIGHT On 29 June 1995, *Atlantis* became the first Shuttle to dock with Mir, the first NASA spacecraft to return to Earth with a partially different crew. *(NASA)*

FAR RIGHT Mir as photographed by the crew of STS-71. *(NASA)*

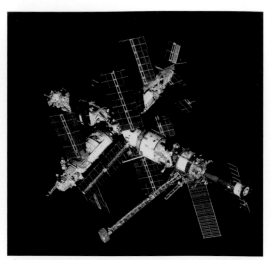

STS-71

27 June–7 July 1995

The first Shuttle to dock with Mir delivered cosmonauts Anatoly Solovyev and Nikolai Budarin and returned ten days later with Thagard, Dezhurov and Strekalov. Thus it fell to *Atlantis* to be the first NASA spacecraft to return with a partially different crew to that which it had carried on launch day and for this mission the resident Shuttle crew going both ways were NASA astronauts Robert 'Hoot' Gibson, Charles Precourt, Ellen Baker, and Bonnie Dunbar. On the approach to Mir, Gibson was at the controls and conducted the approach from below the station, holding formation with Mir at a distance of 250ft (76m) for the vital 'go' from mission control centres in Houston and Moscow. Tension was high as Gibson eased the two together at a speed of 0.7mph (1.13kph), 250 miles (400km) above Lake Baikal during the afternoon local time on 29 June. The docking unit was an androgynous system developed

RIGHT From their Soyuz spacecraft Solovyov and Budarin photograph the Shuttle attached to Mir before it departs for Earth. *(NASA)*

specifically to connect the airlock module in the forward Shuttle cargo bay with the docking port on the Kristall module. Using the Lyappa system, on 10 June Kristall had been moved from its radial position to the forward port on Mir so that *Atlantis* could dock along the long axis of the station rather than off to one side, giving it clearance over solar arrays.

To bring the experience closer to Earthlings, *Atlantis* carried an IMAX camera for cinema-style filming, and five schools in the US had dedicated links to the docked complex for question and answer sessions that would soon become popular on Shuttle flights. Some 15 experiments in human biology and life sciences kept the ten crewmembers busy until undocking on 4 June, prior to which Solovyov and Budarin climbed into their Soyuz spacecraft and separated from Mir to photograph the historic departure before returning to begin a long stay in space. Returning with a record eight crewmembers, *Atlantis* had cleared the way for routine Shuttle–Mir operations and brought astronaut Thagard back from his 115 days in space, extending in duration the 84.5-day US flight record of the third and final Skylab crew, set more than 21 years before.

STS-74

12–20 November 1995

For the second Shuttle–Mir docking mission *Atlantis* carried a new 9,000lb (4,080kg) docking module (DM), 15.4ft (4.7m) tall and 7.2ft (2.2m) in diameter. Development of the DM had been fast-tracked by the need to have a module that could extend the distance between the Shuttle and the Kristall module so as not to interfere with solar cells on the latter, Kristall having been rotated back to its radial docking position on 17 July. To achieve a link-up, the DM had first to be lifted from the rear of the Shuttle's cargo bay to the top of the forward airlock at the front of the bay. In addition to two strap-on solar cell arrays that would be positioned on Mir during a subsequent spacewalk, *Atlantis* also provided 1,000lb (454kg) of water for the Russian station. Shuttle crewmember Chris Hadfield was a Canadian and on 3 September the Soyuz TM-22 spacecraft had delivered cosmonauts Yuri Gidzenko and Sergei Avdeyev and German astronaut Thomas Reiter from the European Space Agency to the station awaiting the arrival of STS-74. Thus Mir hosted crewmembers from all the participating ISS countries with the exception of Japan. When *Atlantis* and its five-man crew separated from Mir on 18 November it left the DM attached to the Kristall module for later Shuttle missions.

STS-76

22–31 March 1996

The third Shuttle–Mir docking delivered Shannon Lucid, the first female NASA astronaut aboard the station, at the start of a planned 142 days in space that would see the first EVA conducted during a Shuttle–Mir docking. The two NASA astronauts assigned to the EVA, Linda Clifford and Rich Godwin, exited via the Shuttle airlock hatch on 27 March to install sets of experiments on the outside of the Mir station and test a variety of tethers, restraints, and special tools that would be used with the ISS. Undocking came on 24 March, Shannon Lucid remaining with cosmonauts Yuri Onufrienko and Yuri Usachev, who were previously launched to Mir aboard Soyuz TM-23 on 21 February. As a historical footnote, this was NASA's last human mission managed from the original Mission Control Center at the Johnson Space Center, Houston. After five days docked to Mir, *Atlantis* separated on 29 March and returned to Earth two days later.

The last of five modules flown to Mir was launched on 23 April 1996, when a Proton

ABOVE From the aft end of the Shuttle cargo bay, a fish eye lens captures *Atlantis* docked to the Mir station on STS-74. *(NASA)*

rocket lifted the 43,400lb (19,685kg) Priroda laboratory into space. It docked to the forward axial port three days later and was transferred to the last remaining side position by the Lyappa system on 27 April. With a length of 42.6ft (13m) and a diameter of 14.3ft (4.36m), Priroda was packed full of equipment not only from Russia but also contributed by Bulgaria,

LEFT Cosmonaut Yuri Onufrienko shows off the interior of the Mir core module. *(NASA)*

BELOW STS-76 delivered Shannon Lucid, the first female cosmonaut to fly aboard Mir. *(NASA)*

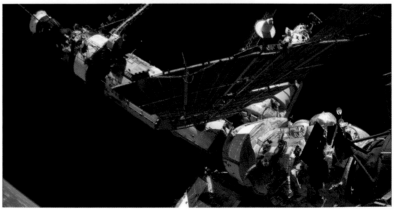

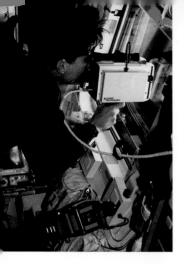

ABOVE Linda Godwin points a laser range finder at the Mir station during rendezvous operations on STS-76. *(NASA)*

ABOVE RIGHT Kvant-1 supports Soyuz TM-23 which had been launched February 21, 1996, and remained with Mir until September 2. *(NASA)*

Germany, Poland, Romania, the Czech Republic, and the USA covering materials science, biotechnology, life sciences, Earth observation, and space technology.

The next Shuttle mission should have launched on 31 July 1996, replacing Shannon Lucid with John Blaha, but that mission was delayed when booster problems kept *Atlantis* grounded. On 17 August Soyuz TM-24 carried cosmonauts Valery Korzun and Aleksandr Kaleri and French astronaut Claudie Andre-Deshays to Mir for a docking, now placing two women aboard Mir for the first time. She returned to Earth on September 2 with Onufrienko and Usachev in the Soyuz T-23 spacecraft, leaving Shannon Lucid with her two new Russian colleagues until *Atlantis* could be cleared to collect her.

STS-79

16–26 September 1996

Heading for the fourth Shuttle–Mir docking, despite delays *Atlantis* finally got off the pad for a ten-day mission with a six-man crew commanded by William Readdy. Docking occurred on 19 September and *Atlantis* undocked four days later having collected Shannon Lucid, depositing John Blaha for his long-duration flight aboard Mir. In space for 188 days – 46 days longer than planned due to Shuttle delays – Lucid, a veteran of four previous Shuttle missions, now had the double record of longest space flight for both an American and a woman. Her record stood for more than ten years until it was broken by Sunita Williams on 16 June 2007, aboard the ISS. *Atlantis* delivered a record 4,600lb (2,087kg) of water, food, and other items to Mir and brought back 2,200lb (1,000kg) of scientific equipment. This was the first flight to carry the 15,500lb (7,000kg) Spacehab double module in the Shuttle's cargo bay, which served as a store for hardware up to Mir and down to Earth. Because Blaha was in orbit during the 1996 US elections he was prevented from casting his vote, a dilemma that a year later resulted in the Texas legislature changing its statutes to permit voting from space!

RIGHT NASA astronaut Shannon Lucid floats in the Mir space station. She was returned to Earth by STS-79 having spent 188 days 4hr 14s in space. *(NASA)*

RIGHT Shannon Lucid moves her space suit back into the Shuttle prior to returning home. *(NASA)*

BELOW STS-79 was the fourth Shuttle mission to dock with Mir, placing John Blaha on board for the third of NASA's long duration stays aboard the station. *(NASA)*

STS-81

12–22 January 1997

Commanded by Michael Baker, the fifth Shuttle flight to Mir offloaded 6,000lb (2,720kg) from *Atlantis* and downloaded 2,400lb (1,090kg) of equipment, returning John Blaha to Earth after his 118-day flight following its five days docked to the Russian station. *Atlantis* delivered Jerry Linenger to Mir at the start of his stay, the third US long-duration mission aboard the station. As on previous visits, valuable working practices, methods, and procedures that would be essential for integrating international crewmembers aboard the upcoming ISS were explored as well as solutions found to unexpected operating difficulties.

Almost three weeks after *Atlantis* returned to Earth, ESA astronaut Reinhold Ewald accompanied cosmonauts Vasili Tsibilev and Aleksandr Lazutkin aboard Soyuz TM-25, launched to Mir on 10 February. The automatic docking on 12 February failed and a manual approach was necessary, after which the three crewmembers joined Korzun, Kaleri and Linenger aboard Mir. But on 23 February a fire broke out when a lithium hydroxide canister exploded. Quickly extinguished, the fire produced toxic gases and the crew wore breathing gear until the fumes dispersed through the air recirculation system.

Then, two days after Korzun, Kaleri, and Ewald returned to Earth in the Soyuz TM-24 spacecraft on 2 March, Progress tanker M-33 was unable to dock and the attempt was abandoned. The Progress spacecraft was a one-way delivery truck and so M-33 had to be sent to a fiery end in the atmosphere on 12 March, packed full of supplies it had been unable to deliver. Finally, on 8 April, Progress M-34 successfully docked at the aft port on Kvant-1 and its supplies offloaded, including a new Elektron oxygen production unit.

STS-84

15–24 May 1997

Mindful of the fire and the failed Progress link-up, it was with some trepidation that *Atlantis* was launched for the sixth Mir docking carrying a crew of seven. British-born astronaut Michael Foale replaced Jerry Linenger for NASA's next long-duration stay aboard the

station. Linenger had logged 132 days in orbit, at that time the longest of any US astronaut, and during his tenure had become the first American to conduct an EVA from a non-US spacecraft wearing a Russian Orlan space suit. Linenger had experienced a near-catastrophic fire, a failure in the carbon dioxide filter system, a failure in the Mir attitude control system and the near collision by a Progress vehicle that failed to dock with the station. And then it got worse.

On 24 June, a month after *Atlantis* returned to Earth, Progress M-34 was undocked for relocation and while manoeuvring around the station it went out of control, crashing into the Spektr module, causing depressurisation and smashing solar arrays. Scurrying to seal off the module, the crew isolated Spektr and saved the rest of the station complex. Three days after the collision batteries on Kvant-2 ran down and TM-25's thrusters were used to re-boost the station's altitude. Progress M-35 arrived on 7 July and docked to the Kvant-2 port, bringing fresh supplies. A week later cosmonaut Tsibilev showed signs of cardiac arrhythmia and his workload had to be drastically reduced.

On 5 August, Soyuz TM-26 carried cosmonauts Anatoly Solovyov and Pavel Vinogradov to Mir where they became Foale's new companions, with the resident crew of Tsibilev and Lazutkin returning to Earth in their TM-25 spacecraft on 14 August. The next day the Mir crew flew around the station in TM-26 to view the damage and assess the condition of the complex. This was followed by a spacewalk conducted by Solovyov and Vinogradov on 22 August to cut Spektr's severed electrical cables and rewire remaining solar panels, restoring 70 per cent of the original capacity. But Spektr itself remained sealed. And then Mir suffered a major computer failure, which was unresolved by the time the crew got it started again. The launch of Progress M-36 was consequently delayed so that a replacement could be carried aloft.

STS-86

25 September–6 October 1997

The seventh Shuttle mission to Mir was the last visit for *Atlantis*, which on this flight delivered 7,000lb (3,175kg) of supplies including 1,700lb (770kg) of water. For five

ABOVE On June 24, 1997, Progress M-34 collided with Mir's Spektr module, seriously damaging its solar arrays. Sealed off, the module was never used again.
(NASA)

days the crew conducted a wide range of experiments, returning with Michael Foale, who had spent 144 days in space, and leaving David Wolff in his place with cosmonauts Solovyov and Vinogradov. Diminutive astronaut Wendy Lawrence was supposed to have replaced Foale aboard Mir but concerns about the minimum size Orlan EVA suit requirement prevented her from flying this mission. For the remainder of the year Mir struggled on with several failures to computers and attitude control systems. Two Progress cargo-tankers delivered supplies and stores as required.

STS-89

22–31 January 1998

The penultimate Shuttle–Mir docking saw *Endeavour* – the replacement for *Challenger*, which was destroyed on 31 January 1986 – make its 12th flight, carrying a crew of seven plus 7,000lb (3,175kg) of stores for the station, docking on 24 January. Astronaut Andrew Thomas replaced David Wolff, who had logged 128 days on his long-duration mission aboard Mir. Undocking came after almost five days of dual activity, Wolff remaining aboard Mir with Solovyov and Vinogradov on what was the last US long-duration visit to the station before translating to the International Space Station, the assembly of which would begin later the same year. For the next four months Mir's resident crew of Talgat Musabeyev and Nikolai Budarin would maintain the station and carry out research through the last Shuttle visit in June. Delivered to Mir after launch by Soyuz TM-27 on 29 January, they came up

with French astronaut Leopold Eyharts, who returned with Solovyov and Vinogradov in Soyuz TM-26 on 19 February.

STS-91

2–12 June 1998

Andrew Thomas had completed 141 days in space when he joined the six-person crew aboard *Discovery* and returned to Earth at the end of the last Shuttle–Mir mission. Commanded by Charles Precourt, STS-91 included in its crew Wendy Lawrence, unable to fly STS-86 due to suit undersize restrictions. The Shuttle had arrived with about 5,800lb (2,630kg) of water and stores and carried a prototype Alpha Magnetic Spectrometer designed to look for dark matter in the universe. Docked for almost four days, *Discovery* departed on 8 June, the station now operated in its last two years solely by Russian cosmonauts in space and controllers on the ground.

Mir made redundant

Russia's Mir space station had been the pride of the Soviet Union but when the communist regime collapsed it was a financial burden the country was no longer able to afford. Had not the agreement been signed with the United States in 1993, Russia would have been unable to continue with Mir and probably would by now be out of the human space flight business. As a result of the agreement, in addition to launch and operational costs, the US paid $472 million for trips to the Russian space station and kept the programme alive. At least, that was, until agreement was finalised on the International Space Station, by which time the economic situation in Russia had improved. In the nine docking missions to Mir the Shuttle delivered 50,500lb (22,900kg) of cargo and equipment and returned to Earth carrying 17,200lb (7,800kg) of redundant equipment and waste, a staggering logistical transfer of 67,700lb (30,700kg). But even at the end, as the Russians kept it alive for two more years, Mir was costing that country $230 million a year just to keep it running. Nevertheless, Mir was a proud icon of a once dominant space programme and the Russians were determined to keep it operational even as they were helping construct the ISS.

After the extensive involvement of NASA and its nine Shuttle docking flights, the Russians struggled to keep Mir operational. The condition of the space station, now more than 12 years old, was declining, and sustainability of what had, in reality, been the world's first international space station was increasingly untenable. Musabeyev and Budarin continued to operate the station until they were replaced by Padalka and Avdeyev in Soyuz TM-28 on 13 August 1998. Many things were changing in Russia and the government divested its interest to a commercial operator known as MirCorp, making it the world's first commercial space platform. Paid for by an agreement between MirCorp and manufacturer Energia, a crew change in February 1999 carried Russia's Viktor Afanaseyev and France's Jean-Pierre Haignere to join Avdeyev while Padalka returned to Earth. When these three crewmembers returned home on 27 August 1999, Mir was left unmanned for the first time in almost ten years. Mir had supported human life for 4,592 days and sustained an unbroken permanent presence for 3,640 days, 22 hours, and 52 minutes since 5 September 1989.

With 60 per cent owned by Energia and 40 per cent by private investors, MirCorp sought to market Mir for space tourists. An early subscriber, American millionaire Dennis Tito, was booked to fly but the Americans resisted giving their approval. The Russians had signed an agreement with the international station partners that they would not mix valuable resources by operating Mir while developing ISS elements. Since STS-91 three Progress tankers had restocked the station but with assembly of the ISS by now well under way, the decision was made to deliberately de-orbit the venerable station. To do that a special Progress vehicle would dock to the Kvant port and, through a series of three rocket burns, bring it down over a remote region of the Pacific Ocean.

Before that, in February 2000 Progress M1-1 docked to Mir and on 6 April a new and improved Soyuz, TM-30, delivered cosmonauts Sergei Zalyotin and Aleksandr Kaleri for a final visit on what was dubbed the 'MirCorp mission'. They conducted a spacewalk, another commercial 'first', on 12 May to examine Mir's exterior and inspect an errant Solar Array Wing on the Kvant-1 module, discovering burnt wires that had cut power. The first 'commercial' Progress cargo-tanker docked with Mir in April and on 15 June Zalyotin and Kaleri returned to Earth. Having a $70 million order backlog for tourist flights, the Russians eventually succumbed to pressure from the Americans – for the time being. Just a few months later the Russians would put Tito aboard the ISS and begin a successful programme of carrying tourists to the station!

The final Progress sent to Mir, M1-5, was launched on 24 January 2001 and docked three days later. On 23 March it performed three de-orbit burns in succession, bringing Mir down to fiery destruction, seen from Fiji at 5.50pm local time. Mir had been in space for 5,519 days and made 86,331 orbits of the Earth at an average altitude of more than 200 miles (322km), travelling a total distance around Earth of more than 2 billion miles (3.22 billion km). It had been an emotional end to a decade of outstanding success. For the first time, a station launched for basic research had grown to many times its original size through the addition of specialised modules and complex equipment lifted from Earth to a permanent home in space. Along the way the station had proven the concept of automated supply and refuelling operations using Progress tankers, derivatives of the manned Soyuz spacecraft first flown in the mid-1960s and now the standard workhorse for taking people to and from low Earth orbit.

In time Soyuz would twice save the future International Space Station from being left untended when first the Shuttle stood down after a catastrophic disaster in 2003 and again after the last Shuttle came home in 2011. But Mir had itself given life to NASA's aspirations for a permanently manned facility of its own, shared with international partners encompassing long-term friends and former adversaries. It is not too extreme a view that had Mir not happened, the ISS would never have been built. Saved politically by a US president, the ISS is the very embodiment of Cold War attitudes applied to post-Cold War logic, for a cohesive and cooperative bond between technologically superior states.

Chapter Three

ISS Phase Two – assembly

What began as an idea to work together in space, assembling a giant orbiting laboratory, developed only gradually as technical and political differences intervened to slow the process. The big challenge was learning how to trust each other, former adversaries working as one for the first time.

OPPOSITE Sunrise over Zvezda, seen 16 times in one Earth day. *(NASA)*

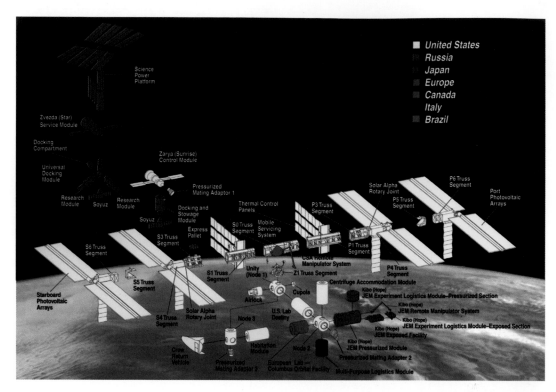

RIGHT The various elements of the ISS colour-coded according to the country of origin, looking more like a child's construction kit. *(NASA)*

By 1995 the scale of the ISS had grown to 75 assembly flights in a complex schedule of launches planned to start in November 1997 and finish in June 2002. By 1997 the number of prime flights had been cut back to 45 and the final configuration settled. The partners finally agreed on a 'T' configuration for the ISS modules, with a single truss assembly supporting eight individually geared solar array panels, four at each end. The long axis of the 'T' modules would have a length of 167ft (51m) and the top bar of the 'T' would represent the opposing reach of the European Columbus and Japanese JEM modules located either side of the Node 2 (Harmony) module. The truss supporting the four Solar Array Wings at each end would be attached to the top of the US module and have a span of 357ft (109m). It would take more than a decade to build the ISS and because of that it would evolve incrementally, with several elements moved around to balance the geometry of the configuration but also make use of each element in a productive way. Because of this, the location of individual cargo elements would change as the station grew, complicating a straightforward explanation of how it was assembled.

A key part of the design was the need for the structure to be stable at every step in the assembly process. Although weightless, any

structure in orbit has mass and inertia and the Earth's gravity tends to pull it around to point downwards like a pendulum. Also, the layout of the structure must be arranged around its centre of mass so that aligning it correctly for operational use will not pose attitude control problems. It was these unseen forces playing on all structures in space that bedevilled early configurations and experience with controlling massive objects such as NASA's Skylab and Russia's Mir were useful in tweaking the design and assembly sequence for the ISS. Another consideration arose from the length of time needed to completely assemble all the large and complex truss beams for supporting the huge solar arrays. Some of those arrays would be needed early on in the assembly phase to provide electrical power for the station and would have to be moved around to convenient locations until the final main truss structure had been assembled.

Under the new plan, the Shuttle–Mir flights of Phase 1 would be followed by Phase 2, beginning with the launch of the first element and ending with sufficient assembly to begin permanent habitation. By 1998 it was hoped to complete Phase 2 by early 2000. Phase 3 would mark completion of full assembly with all modules and structures installed, anticipated for 2004. The first element launched was the Russian-built Zarya which would be followed by a Shuttle carrying the first of three nodes called Unity, docked to one end, followed by another Russian-built module called Zvezda docked to Zarya. Together they would comprise the Russian section of the ISS. Docking ports on the Russian section would allow other modules to be attached after assembly had been completed.

Zarya/ISS 1AR

20 November 1998, 6.40am GMT

Zarya's launch was delayed beyond November 1997 when construction fell behind schedule due to financial difficulties on the Russian side. Under the original plan put together in 1994, Boeing would build the first element of the station known as Bus-1 for a cost to the US taxpayer of some $450 million. As the price of Bus-1 began to rise still higher, Boeing, under pressure from NASA, was persuaded to cancel it and pay a Russian manufacturer, the Khrunichev State Research & Production Space Centre (KhSc), to deliver a replacement named Zarya ('sunrise' in Russian), which had been the name Russian engineers had wanted to give the Salyut 1 space station almost three decades earlier.

Zarya was based on the so-called Functional Cargo Block (FGB), and was bought by the Americans for $220 million, the Russians having written off its development costs in those earlier programmes. Designed essentially for living and working in, the FGB evolved from the TKS resupply ship for the Russian Almaz military space station programme. That had been first launched as Cosmos 929 on 17 July 1997. That design formed the basis on which Mir modules were developed and was now to become the first ISS element launched into space.

Assembly of Zarya began in December 1994 and it was delivered to the Baikonur cosmodrome in January 1998 as the political wrangling over delays reached fever pitch. With problems emerging at this early stage many US politicians feared the worst, citing Russian corruption and obfuscation as reasons for America to pull out of the programme. Political pressure was applied and the funds that should have been going to produce Zvezda were

LEFT A Proton-K rocket launches the Zarya ISS module on 20 November, 1998. (NASA)

found. Zarya was launched by Proton rocket on 20 November 1998, the first element in an assembly sequence that would last more than 12 years.

Anatomy
Zarya module

Length	41.2ft (12.56m)
Diameter	13.5ft (4.11m)
Pressurised volume	2,525ft^3 (71.5m^3)
Solar array span	80ft (24.39m)
Weight	42,600lb (19,323kg) fuelled and 17,600lb (7,983kg) unfuelled
Launched	20 November 1998
Launch vehicle	Proton
Launch site	Baikonur

Visually similar to the Mir core module, Zarya had an internal pressurised compartment and a rear equipment and propulsion section and would serve initially as a powerhouse and tug. It was equipped with a Kurs automatic rendezvous and docking system and the forward section comprised a spherical compartment with single forward (axial) and radial docking ports to which Soyuz and

RIGHT STS-88
launched on
December 15, 1998,
and approached Zarya
for a docking 60 hours
47 minutes later.
(NASA)

Progress vehicles could be attached. Zarya was equipped with 24 docking and stabilisation motors, each with 88lb (40kg) thrust, and 12 small, 2.9lb (1.3kg) thrust stabilisation motors. Two rocket motors for orbital changes essential to rendezvous and docking each had a thrust of 919lb (417kg). The 12,698lb (5,760kg) of N2O4/UDMH propellant was carried in external tanks, with eight short tanks each holding 105.6 gallons (480 litres) and eight long tanks each with 87.17 gallons (396 litres). Up to 3.3kW of electrical power was produced by two solar arrays, each 35ft (10.7m) long and 11ft (3.35m) wide, and the module would provide orientation and propulsion for the early configuration of ISS during the initial stages of assembly.

In cooperation with Boeing, Honeywell installed multiplexer/demultiplexer units enhanced with computer software and Intel 80386 processing cards. The pressurised compartment contained equipment for life support, command and data handling, electric power, and docking. For attitude control several star, sun and horizon sensors provided data integrated with measurements from magnetometers, rate gyroscopes and Russia's Glonass navigation satellites. Commands from Russia's mission control could be sent through the Regul system or from Luch satellites. Inside the module, analogue communications provided voice, paging, and caution and warning signals with information received and transmitted via two audio communication units operating at VHF frequencies.

To provide electrical power the two solar arrays unfold in orbit. Each array supports 14 solar cell panels and comprises six electronically independent generators, each with 85 solar cells connected in series. Output varies from 24–34 volts. For two-thirds of each orbit Zarya would be in the Earth's shadow and the arrays would be motionless. Power to the module at these times would be provided by a battery that stores electrical energy when the arrays are in sunlight. To charge the battery and provide direct power the arrays track the sun as it arcs through the sky as the module orbits the Earth every 90 minutes. The maximum allowable power (cells plus battery) is 13kW on the sunlit portion of the orbit and 6kW in shadow.

The Environmental Control and Life Support System (ECLSS) maintains a sea-level atmospheric pressure of 14.7lb/sq in, maintaining a mixed gas atmosphere of oxygen and nitrogen. The nitrogen is manually controlled but oxygen is supplied by Russia's Elektron water electrolysis unit. This device collects and condenses moisture from humidity in the atmosphere and waste water which is broken down into hydrogen and oxygen using electrolysis. Hydrogen is vented to the vacuum of space and the oxygen is used to mix with the nitrogen to produce a breathable atmosphere. The system automatically maintains a constant output which, if it varies, is regulated by a change in current. This was new for Americans, who had never recycled atmospheric products in any of their spacecraft. The Russians had used the system in Mir, but with varying and sometimes unreliable results.

Water production is based on an average consumption of 20–25 litres per person per hour. The ECLSS also controls circulating fans, heat exchangers, fire detection equipment, and a gas analyser which detects pollutants. Thermal control is maintained by means of an ethylene-glycol and water mixture through two loops, one of which is a back-up. The loop's heat bearing fluids pass through 12 externally mounted radiator panels containing pipes filled with ammonia, transferring thermal energy to heat pipes and thence to space through radiation.

ABOVE The mission patch for the first Shuttle flight to the Zarya module, STS-88. *(NASA)*

STS-88/ISS 2A Unity

Endeavour
4 December 1998, 8.36am GMT
Cdr: Robert D. Cabana (3)
Pilot: Frederick W. Sturckow
MS1: Jerry L. Ross (5)
MS2: Nancy J. Currie (2)
MS3: James H. Newman (2)
MS4: Sergei Krikalev (1) (Russia)

The second ISS element lifted into space was the first of three nodes. Called Unity, it was the first piece of US-built hardware lifted to the nascent station but the node was one of the most important components of the ISS because it had unique responsibilities. When NASA downsized the station from four primary US modules to two in an effort to cut costs it needed somewhere to put all the systems equipment essential to running the facility. So the concept of the node emerged, places where engineering and systems equipment could be located while also doubling as spacers between modules and other structures. Three nodes were built, one assembled by Boeing in the US (Unity) and two slightly bigger nodes (2 and 3) manufactured by agreement with the European Space Agency.

The launch of Node 1 and its two Pressurised Mating Adapters took place when Shuttle *Endeavour* lifted a crew of six from the Kennedy Space Center to an initial orbit of 175 x 87.2nm.

ABOVE Carrying the second ISS element into space, *Endeavour* lifts the Unity Node 1 module from the interior of the cargo bay. *(NASA)*

LEFT Installing Unity was a delicate job, the inertia of the 25,000lb structure controlled by precision movement using the Shuttle's Canadian-built manipulator. *(NASA)*

LEFT PMA-1 attached to Unity is gently lowered to the Docking Module on the Shuttle *Endeavour*. *(NASA)*

ABOVE A fish-eye lens distorts the view of the Unity and its two PMAs attached to the Shuttle Docking Module and the Shuttle manipulator arm grasping Zarya. *(NASA)*

Through a series of manoeuvres *Endeavour* narrowed the gap to Zarya, 208nm above the Earth. On the second day after launch the Shuttle's remote manipulator arm grappled Node 1 and its two attached Pressurised Mating Adapters and lifted it from its cradle, moved it through 90 degrees and connected it to the Orbiter docking system at the forward end of the payload bay at a mission elapsed time of 39hr 9min.

Several additional rendezvous manoeuvres

were performed before *Endeavour* was at Zarya, about 60hrs 47min into the mission. The Shuttle manipulator arm was raised from its stowed position and at 63hrs 12min attached itself to the grapple fixture on Zarya and at 65hr 31min into the mission *Endeavour*'s thrusters were blipped to dock Unity with the Russian module. In Houston it was 8.07pm on 6 December and in Moscow it was 5.07am on 7 December. The first connection between two elements of the International Space Station had been achieved, creating a structure 76ft (23.2m) long and weighing over 31 tonnes. But a lot of work and two spacewalks were needed before the first human occupants could move from *Endeavour* to Unity and on into Zarya.

Jerry Ross and Jim Newman performed a 7hr 21min EVA on 7 December by leaving through the airlock beneath the docking interface with PMA-2 and hooking up 40 connections and cables. Activation sent power surging into Unity at an elapsed time of 91hr 13min. During the spacewalk they inspected the exterior of both Unity and Zarya and examined an antenna that had failed to deploy after launch. The next day the Shuttle's thrusters fired 11 times over a period of 21min 47sec to raise the combined orbit of the docked vehicles to 215 x 211nm. On 9 December a second EVA lasted 7hr 2min during which Ross and Newman removed restraint pins from the Common Berthing Mechanism on Unity, installed covers on two data relay buses, freed a back-up antenna on Zarya, and installed two S-band antennas on Unity. Only now could they prepare to open up Unity and float on through from the Shuttle's docking module.

Cabana and Krikalev floated into Node 1, side by side to symbolise partnership, at 155hr 18min elapsed time. In Houston it was 1.54pm on 10 December. Just 1hr 20min later, after bear hugs and a handshake, they opened the hatch into Zarya and floated across. In Houston ISS manager Randy Brinkley waxed unusually lyrical while Boeing Zarya manager Virginia Barnes summed it up, affirming that 'what happened today...signified not only tremendous engineering but also an emotional journey.' The next day, 11 December, the crew transferred 1,200lb (544kg) of equipment from *Endeavour* into Unity, stowing 335lb (152kg) of equipment from Zarya to the Shuttle. Turning

RIGHT Astronauts James Newman and Robert Cabana check out the Unity module. *(NASA)*

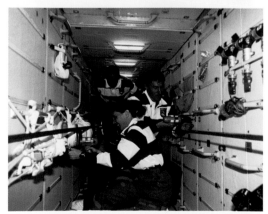

RIGHT Inside Zarya, astronauts reconfigure the equipment from a launch condition to an operational environment. *(NASA)*

out the lights after 28hr 32min aboard the first two elements of the ISS, the crew withdrew into *Endeavour* and prepared for a final EVA.

The final spacewalk took place on 12 December when Ross and Newman freed an antenna on Zarya, leaving a tool strapped close by for future spacewalkers to use should they need it, and discovered that an experiment tray on the outside of the Russian module had a loose door flapping open. Moving more than 70ft (21m) above the Shuttle to the far aft end of Zarya, they marvelled at the view below them as they installed a new handrail that could be attached there before launch. Before returning through the airlock into *Endeavour* Ross tested a small emergency backpack called a SAFER (Simplified Aid For EVA Rescue) equipped with tiny nitrogen jets as thrusters, a device first tested in 1997 aboard Mir but which on that occasion had failed. The 6hr 59min EVA brought to an end the operational start to an assembly process that would last more than 12 years.

With PMA-2 depressurised, on 13 December the Shuttle and its docking module slipped away from the docked station at an elapsed time of 227hr 49min and began a fly-around inspection of the station. They had been docked for 6 days and 18 hours. After a final burn to start *Endeavour* drifting away from the ISS, preparations for re-entry began in earnest. The following day the crew released a small US Air Force satellite, MightySat, that had been carried in the payload bay, and a small satellite for Argentina. *Endeavour* landed back at the Kennedy Space Center on 15 December at an elapsed time of 11 days, 19 hours and 18 minutes.

ABOVE LEFT Jerry Ross conducts work outside Unity fixing cables and attaching conduits between Node 1 and Zarya. *(NASA)*

ABOVE Unity (foreground) attached to Zarya as viewed by a space walking astronaut. *(NASA)*

BELOW As *Endeavour* moved away its crew viewed the first unified building block for the ISS – Zarya and Unity. *(NASA)*

RIGHT An engineering
drawing shows the
two Pressurised
Mating Adapter units
at opposing ends
of the Unity Node 1
module. *(NASA)*

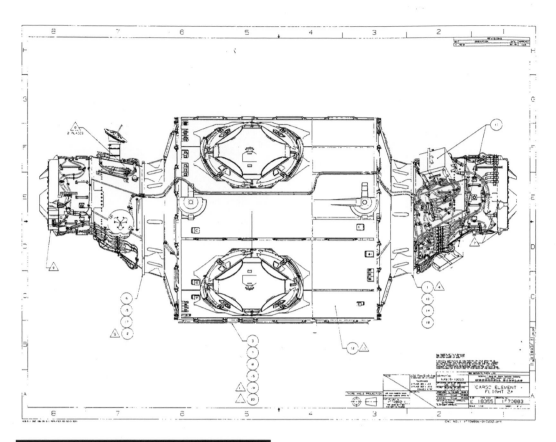

Anatomy

Node 1: Unity module/PMA-1 & 2

Length	18ft (5.5m) and 34ft (10.4m) including 2 PMAs
Diameter	15ft (4.6m)
Weight	25,600lb (11,600kg)
Launched	15 December 1998
Launch vehicle	Shuttle *Endeavour* STS-88
Launch site	Kennedy Space Center

Unity has six docking ports, one axial port at each end on the half-cone endplates and four radial ports placed at 90-degree intervals around the cylindrical structure. The six ports each have a Common Berthing Mechanism (CBM), a standard design of access and hatch used to link together all the non-Russian pressure modules throughout the ISS. They comprise two rings, an active CBM (ACBM) and a passive CBM (PCBM), performing the relevant functions their names imply and when connected together the two halves of the CBM form a pressure-tight seal. Each CBM incorporates a square hatch which when opened provides a 50-inch (1.27m) clearance

for moving equipment between modules. The six docking ports on Unity are of the ACBM (active) type.

Unity was launched with forward and aft CBMs, each supporting a conical Pressurised Mating Adapter (PMA) about 8ft (2.44m) long. Each PMA comprised a truncated ring-stiffened shell structure machined from 2,219 aluminium roll-ring forgings welded together. With a 28-in (71cm) axial offset, all three PMAs used by the ISS are identical, acting as an interface between vehicles to which each was docked. Each PMA was an independent segment capable of being removed from Unity but they were preassembled with Unity for launch and on-orbit connection to other elements. As such, they constituted adapter units with a hatch at each end permitting pressure equalisation checks before opening the CBM into Unity. PMA-1 weighed 3,504lb (1,560kg) and joined Unity to Zarya at the aft end. Weighing 3,033lb (1,375kg), PMA-2 joined Unity to the Orbiter at the forward end, initially enabling Unity to receive Shuttle visits.

Each PMA carried hybrid computer multiplexer-demultiplexer units mounted externally. Pressurised internally and with handholds on the inside, initially they provided internal passageways into and out of the Unity node. PMA-1 would remain as the connection between the US and Russian sections for the life of the ISS but at various times during the assembly sequence PMA-2 would be moved to different locations, eventually taking up permanent residence at the top of the Harmony module (Node 2). PMA-3 would be delivered by STS-92 in October 2000 but each would provide primary docking ports for the Shuttle.

During the STS-116 mission of December 2006, PMA-2 would be fitted with the Station–Shuttle Power Transfer System (SSPTS), which served to provide power to the Orbiter while it was docked to the station. Incorporating a Power Transfer Unit (PTU) which replaced the Assembly Power Converter (APCU), it provided the ability to convert the 120 volt DC power from the ISS solar arrays to the Shuttle's 28 volt DC main bus. Up to 8kW could be fed from the ISS to the Shuttle, relieving demand from the Orbiter's fuel cells and adding up to four days docked at the ISS. The first operational use of this system came with the STS-118 mission in August 2007 but only *Discovery* and *Endeavour* were so modified, *Atlantis* being constrained to shorter visits.

An Androgynous Peripheral Assembly System (APAS) was attached to Zarya to connect it to PMA-1. An APAS consists of a structural ring, a moveable ring, various alignment guides, latches, hooks, and fixtures to mate with a copy of itself, each being either active or passive, hence 'androgynous'. A passive attach system was used on the forward end of PMA-2 to mate it to the Shuttle-docking module in the forward area of the cargo bay. During mating, the active half of the APAS capture ring is extended outwards from the structural ring towards the passive half and captures it, while an attenuation mechanism damps out relative movement between the two structures. With the two vehicles aligned, the capture ring is retracted inside the structural ring, whereupon 24 hooks lock the connection together to form an airtight seal. Two APASs were carried, one attached to each PMA.

Unity is equipped with four International Standard Payload Racks, or ISPRs, located at

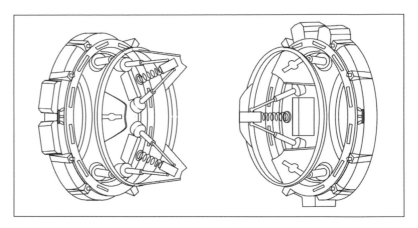

ABOVE The APAS docking system used to connect the PMA to Zarya. *(NASA)*

LEFT The 'active' side of the APAS unit, shown here mounted to the Shuttle Docking module. *(NASA)*

90-degree intervals around the periphery of the docking port that connects it to the Russian sector of the station, but it also doubles as an engine room with systems and equipment essential to the early integration and operational use of the evolving station. Fabricated from aluminium, Unity carried electrical power initially supplied by Zarya and distributed to other modules when they were attached, providing fault protection to individual branch lines. Items

LEFT The 'passive' side of the APAS unit attached to PMA-2 which would be used for docking to the Shuttle, PMA-1 being the permanent fixture between Unity and Zarya. *(NASA)*

ABOVE Unity was launched with equipment stowed against the walls that would be unstowed in space and deployed in permanent locations. *(NASA)*

so attached included fans for air circulation, internal lighting, emergency egress battery chargers, communication systems, heaters, and facilities for future equipment installed on later visits. Power was supplied to Unity via cables routed through PMA-1, including umbilicals for secondary power, one for a data bus and three for power transfer. Six of the power umbilicals were attached during EVA operations conducted by the STS-88 crew of which two were for primary power and four for truss segments installed on later missions. Node 1 computers worked in unison with Zarya's computers to provide all ISS command and data handling services routed through three buses and a further three connected on later missions.

Thermal control was initially passive and then was provided by the Early External Active Thermal Control System (EEATCS) and finally by the External Active Thermal Control System (EATCS). Passive control involved shell-mounted patch heaters and multilayer insulation with Zarya supplying the electrical power for heaters. The EEATCS was installed with the P6 solar array package carried to the ISS by *Endeavour* on STS-97 in November 2000. These transferred heat using liquid ammonia in two identical loops operating at 35–41°F (2–5°C), each loop connected to an interface heat exchanger. The EATCS was activated after the assembly of the P3 and P4 trusses and associated solar arrays brought to the ISS by *Atlantis* on STS-115 in September 2006. The Internal Active Thermal Control System (IATCS) comprised the low, moderate and high-temperature thermal transport

loops. Because Unity was launched dry, these loops previously filled with nitrogen would only take on water when that was delivered by the Multi-Purpose Logistics Module (MPLM) Leonardo prior to the US laboratory module Destiny which brought the Fluid Systems servicer.

The Environmental Control & Life Support System (ECLSS) was designed to maintain a habitable environment and carried inter-module ventilation equipment, pressure sensors, and sundry ECLSS support equipment. The Atmosphere Control and Resupply system monitored the atmosphere and total pressure, measuring oxygen partial pressure, providing nitrogen and oxygen to the atmosphere and equalising pressure between adjacent modules. The Atmosphere Revitalisation section provided oxygen regeneration and removal of carbon dioxide, monitoring trace contaminants and hazardous atmospheres. Temperature and Humidity Control was responsible for removing moisture and heat from the air by circulating and ventilating the atmosphere and removing microbial airborne contaminants through filters, inhibiting 99.97 per cent of particles 0.3 microns or larger. The Fire Detection and Suppression system includes smoke detectors, a caution and warning panel, gas masks, and oxygen bottles and portable fire extinguishers. Unity carried 50,000 mechanical components, with 216 fluid and gas lines plus 121 internal and external cables running 35,000ft (10,668m) of electrical wiring.

Unity carried two multiplexer/demultiplexers running application software and process information, functions which are not normally carried in the multi-channel, simultaneous message transmission and de-channelling capabilities provided by such systems. Data and commands are exchanged via Mil-Std 1553B buses with an Intel 80386SX chip forming the base for the main processing card. The computers provide early command and control of Node 1 and were used for electrical functions and systems management including that of certain photo-voltaic control units. The computers are mounted on the exterior of PMA-1 on chassis platforms designed to protect them from debris and radiation. The computer cases are designed to slowly leak and equalise with their exterior environment, allowing them to be moved in and out of Unity if necessary.

STS-96/ISS 2A.1

Discovery
27 May 1999, 10.49am GMT

Cdr: Kent Rominger (3)
Pilot: Rick D. Husband
MS1: Tamara E. Jernigan (4)
MS2: Ellen Ochoa (2)
MS3: Daniel T. Barry (1)
MS4: Julie Payette (Canada)
MS5: Valery Tokarev (Russia)

The second flight to the ISS included a Russian cosmonaut and a Canadian astronaut on a logistics and resupply flight to keep Zarya and Unity up and running until the much delayed Zvezda could be launched. More than 3,600lb (1,630kg) of equipment in 750 separate items was uplifted by *Discovery*. Most of this equipment was housed within a 16,072lb (7,290kg) pressurised Spacehab double module carried at the rear of the cargo bay, access to which was gained by the astronauts along a tubular tunnel stretching from the Orbiter middeck the length of the cargo bay. The Spacehab double module is a 17ft (5.2m) long structure, 14ft (4.3m) wide and with a height of 11.2ft (3.4m). It had an internal volume of 1,100ft^3 (31.15m^3) capable of carrying 61 lockers or experiment racks, each 2ft^3 (0.057m^3) in volume. Developed for Phase I Mir missions, the double module would be used periodically during ISS build-up. In addition, mounted in the cargo bay above the access tunnel was a flatbed pallet and keel yoke assembly known as the Integrated Cargo Carrier (ICC).

Hail damage to the external tank while the Shuttle was on the pad caused a seven-day delay in the flight. Launch came at 6.50am local time on 27 May 1999, and *Discovery* was placed in an 11 x 183nm orbit, docking to PMA-2 on the forward end of Unity at an elapsed time of 41hr 48min. Just over two hours later the hatch to PMA-2 was open and the crew began to move inside Unity and Zarya, opening it up after a period of more than five months unoccupied. The single scheduled EVA of the mission began after two days and 16 hours of elapsed time from the Shuttle airlock, and, during a spacewalk lasting 7hr

LEFT In May 1999 *Discovery* revisits the Zarya/Unity combination. *(NASA)*

55min, Jernigan and Barry transferred two folded cranes to the exterior of the station, set up new foot restraints, and installed tool bags and handholds for future assembly work. In all, 661lb (300kg) of cargo was transferred to the outside of the station.

About 13 hours after the conclusion of the EVA the Spacehab storage module was activated and the lengthy job of transferring all the equipment into the ISS began. A total of 2,881lb (1,307kg) in 98 items of dry cargo plus 686lb (311kg) of water with 197lb (89kg)

LEFT Tamara Jernigan moves a part of the Russian Strela crane during a space walk on May 30, 1999, during the STS-96 mission. *(NASA)*

of cargo in 18 items moved into *Discovery* to bring home. After several days of activity, moving equipment, conducting experiments, and reconfiguring Unity and Zarya for the next phase, the Shuttle's thrusters were used to re-boost the altitude of the station. After a total of 5 days, 18 hours and 15 minutes docked together, *Discovery* separated from PMA-2 and drifted away, returning to Earth after 9 days, 19 hours and 13 minutes in space.

Anatomy
Integrated Cargo Carrier
Weighing 3,050lb (1,383kg) and constructed of aluminium, the ICC is 13ft (4m) wide across the bay and 8ft (2.44m) long with a depth of 10in (25cm). It incorporates a keel yoke for securing it to the bottom of the fuselage middeck, the lower part of the Orbiter cargo bay. The ICC is never removed during flight and serves effectively as a flatbed pallet to which can be attached up to 8,000lb (3,630kg) of cargo. The ICC was a preferred location of ISS Orbital Replacement Units (ORUs) carried to locations outside the ISS on modules and truss assemblies. For its first flight aboard STS-96 it carried 190lb (86kg) of parts for the Russian Strela crane, stowed on PMA-2 during a spacewalk, and 400lb (181kg) of EVA tools and flight equipment to assist crewmembers during assembly of the ISS.

The ICC was flown on STS-96, STS-101, STS-102, STS-105, STS-106, STS-108, STS-114, STS-121, STS-116, STS-122, STS-126, STS-127, STS-128, STS-131, STS-132, and STS-135. On STS-122 the version flown was known as ICC-Lite, being 4ft (1.2m) long rather than 8ft (2.4m) due to the small size of its load. Two flights, STS-127 and STS-132, carried an ICC-VLD (Vertical Light Deployable) version which differed in having heating and electrical power connections and being removable from the cargo bay using the Shuttle's robotic arm (see STS-127 payload description). Its empty weight was 2,645lb (1,200kg). Seven LCC missions (STS-108, STS-114, STS-121, STS-126, STS-128, STS-131, and STS-135) employed the Light Multi-Purpose Experiment Support Structure Carrier (LMC) which was a lightweight (946lb/429kg) cross-beam version without the keel yoke.

STS-101/ISS 2A.2a

Atlantis
19 May 2000, 10.11am GMT
Cdr: James D. Halsell (4)
Pilot: Scott J. Horowitz (2)
MS1: Mary Ellen Weber (1)
MS2: Jeffrey N. Williams
MS3: James S. Voss (3)
MS4: Susan J. Helms (3)
MS5: Yuri Usachev (Russia)

The launch of *Atlantis* was delayed for more than a month by bad weather and technical problems and was a logistics and resupply mission similar to that flown by *Discovery* a year earlier, carrying a double Spacehab module and an Integrated Cargo Carrier. The two docked modules had been unattended in space for a year, a necessity driven by extensive delays to the Russian Zvezda module which should have flown early in 1999. Held back by financial and operational problems, Zvezda was vital as the next module in the assembly sequence. But less than two months prior to the launch of *Atlantis* the Russians had finally bid farewell to Mir and brought it back to destruction in the Earth's atmosphere. Now they could focus on the ISS and cosmonaut Yuri Usachev, a veteran of 376 days aboard Mir and six spacewalks, accompanied five Americans when *Atlantis* was launched in May 2000.

Atlantis docked to PMA-2 at 1day 18hr 33min. But before entering the ISS, Voss and Williams performed a 6hr 44min spacewalk to transfer 372lb (169kg) of equipment outside including additional parts for the Strela crane which was moved to its permanent stowage location, more handholds, and tools. They also needed to perform several tasks setting up the exterior for the next missions. After the EVA the crew opened up the ISS modules and began logistics transfer and 'housekeeping' duties. The Spacehab double module was opened and 2,657lb (1,205kg) of dry cargo plus 187lb (85kg) of water in four contingency containers was moved to the ISS, with 1,291lb (586kg) moved into *Atlantis*. Extensive repair work on four of the six Zarya batteries turned astronauts into troubleshooters but technical exchanges between the ISS, Houston, and Moscow

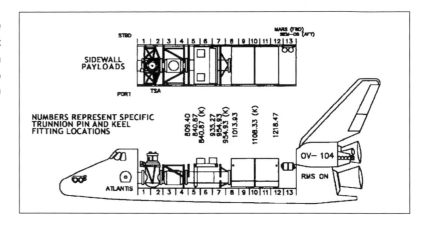

tracked down the problems and a decision was made to replace two on the next mission.

A series of re-boost manoeuvres conducted by the Shuttle raised the docked vehicles to an orbit of 206.7 x 199.5nm, an altitude increase of almost 10nm. The mission had been extended by one day to accommodate the battery troubleshooting and several other minor tasks that built up on the timeline and when it undocked at an elapsed time of 7 days 12hr 52min, *Atlantis* had been moored to the ISS for 5 days 18hr 19min. Engineers examining hours of high-speed imagery of the launch discovered a flash of light under the port wing indicative of damage to thermal protection tiles on the Shuttle, so a special type of re-entry was flown to minimise heating in that area. Nevertheless, when *Atlantis* landed it displayed several heavily damaged tiles in other places and in one the structure of the vehicle itself had almost burned through, a nightmare that would re-visit NASA less than three years later.

Zvezda/ISS 1R

12 July 2000, 4.56am GMT

Consisting of the Dos-8 core module designed for Mir-2, the Zvezda spacecraft is known as the Service Module, the first of the Russian-owned modules launched to the ISS. Zvezda means 'star' in Russian but, as related earlier, this module was reluctant to shine, with delays to its manufacture holding back construction of the ISS by a year and then delaying it further through postponed launch dates. Launched more than two years behind its original schedule, Zvezda was lifted into orbit by Russia's powerful Proton-K launch vehicle. Once separated from Proton, Zvezda used its propulsion systems to adjust the orbit for a rendezvous with the two on-orbit elements of the ISS.

Using the Kurs rendezvous system, Zvezda completed the approach phase and

ABOVE The Zvezda service module during manufacture and assembly prior to launch. *(RKK)*

LEFT Following long delays, Russia's Zvezda module is launch by a Proton rocket in July 2000, the third ISS element. *(NASA)*

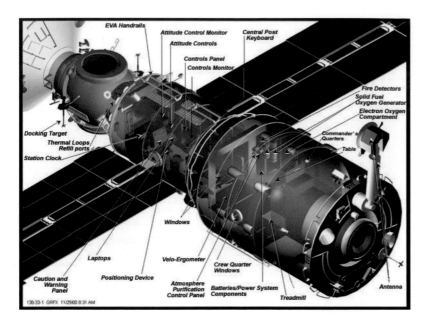

EVA Handrails
Attitude Control Monitor
Attitude Controls
Central Post Keyboard
Controls Panel
Controls Monitor
Fire Detectors
Solid Fuel Oxygen Generator
Electron Oxygen Compartment
Docking Target
Thermal Loops Refill ports
Station Clock
Commander's Quarters
Table
Windows
Laptops
Velo-Ergometer
Crew Quarter Windows
Caution and Warning Panel
Positioning Device
Atmosphere Purification Control Panel
Batteries/Power System Components
Treadmill
Antenna
136-33-1 GRFX 11/29/00 8:31 AM

Anatomy
Zvezda

Length	43ft (13.1m)
Diameter	13.5ft (4.1m)
Pressurised volume	2,649ft^3 (75m^3)
Habitable volume	1,649ft^3 (46.7m^3)
Solar array span	97.5ft (29.7m)
Weight	42,000lb (19,050kg)
Launched	20 July 2000
Launch vehicle	Proton
Launch site	Baikonur

The Zvezda module had a long pedigree dating back to the 1980s and even the structural layout is nearly identical to the core module of the Mir space station. Zvezda comprised four primary sections: the Transfer Compartment, the Work Compartment, the Transfer Chamber, and the Assembly Compartment. The spherical Transfer Compartment is at the front and provides the main interface with Zarya attached at its main axial docking port. Unlike the Mir core module, Zvezda has only two additional docking ports in this compartment – one up and one down – and unlike Mir it has no Lyappa system for transferring modules. These ports were designed to receive Soyuz and Progress spacecraft but other small modules have been attached, including the Pirs docking compartment and the Mini-Research Module-2. Pirs is like the docking module attached to Mir and allows two astronauts to exit Zvezda without depressurisation. Zvezda was built to accommodate three crewmembers, or six in a transfer function.

ABOVE Zvezda was the second major Russian element added to the station, greatly increasing its capacity and adding electrical power. *(NASA)*

then became the passive vehicle while Zarya accomplished the docking. The date was 26 July in Moscow and 25 July in the US. Now comprising three elements, the Zvezda–Zarya–Unity assembly had a total length of almost 120ft (36.5m) and a mass of 110,000lb (49,900kg). For six weeks the docked assembly would remain unattended, awaiting the arrival of *Atlantis* on its second visit to the ISS that year.

Before that, on 8 August 2000, the first unmanned cargo-tanker vehicle sent to the ISS docked to the aft port on Zvezda. Launched two days earlier, Progress M1-3 carried supplies to the station and was ready for the crew of *Atlantis* when it arrived a month later. M1-3 would remain docked to the aft Zvezda port for 84 days until de-orbited on November 1.

RIGHT With a pedigree back to the 1980s, Zvezda owes much to the Mir core module design and is functionally divided into separate working areas. *(NASA)*

RIGHT Cosmonaut Yuri Usachev in Zvezda working a laptop. *(NASA)*

The Transfer Compartment consists of a small, spherical volume with a forward docking port to engage the aft end of Zarya, a zenith (upper) docking ring on top, and a nadir (lower) docking ring at the bottom. Aft of the Transfer Compartment, the cylindrical main Work Compartment is divided into a forward section 9.5ft (2.9m) in diameter and an aft section 13.5ft (4.1m) in diameter providing most of the living and work space in the Russian sector of the ISS. The forward section contains engineering panels, caution and warning displays, lighting control panels, maintenance equipment, and a body mass measurement device.

The aft section contains the environmental control equipment and waste management (toilet) systems together with two crew compartments for individual astronauts. Eventually, it was also equipped with a US Treadmill & Vibration Isolation System (TVIS) and the Russian Vela Ergometer for exercise and physical fitness monitoring. The Work Compartment also provides a kitchen area with a refrigerator-freezer and a table platform for securing meals in the weightless environment. Altogether, Zvezda has 13 optical ports including a 9in (23cm) diameter window in the Transfer Compartment for viewing docking operations, a large 16in (41cm) window in the Work Compartment, and a single window in each crew compartment, which double as sleep stations. Other windows are situated at various locations for viewing the Earth and other modules.

At the extreme aft end of Zvezda is the Transfer Chamber, essentially a short section through which crewmembers can pass from the Work Compartment to the aft docking port, the fourth with which Zvezda is equipped. Wrapped around this tunnel is the Assembly Compartment, comprising the aft end of the module, which contains the propellant tanks and rocket motors for orbit adjustment and attitude control. The Zvezda propulsion systems use the same N2O4/UDMH propellant combination as Zarya. Mounted on the aft face of the module, the two main engines each have a thrust of 661lb (300kg) and can be fired individually or in pairs for orbit adjustment and manoeuvring. In addition, there are two rings of 16 redundant thrusters, each delivering a force

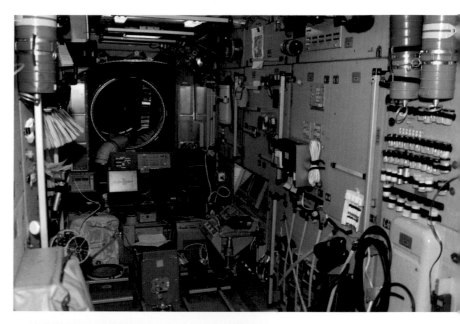

ABOVE A view looking forward toward the spherical Zvezda Transfer Compartment. (NASA)

LEFT As assembly progressed and logistical support increased, Zvezda became increasingly crowded. Krikalev floats through the main work compartment. (NASA)

LEFT Sergei Krikalev watches *Atlantis* through one of the windows in Zvezda. (NASA)

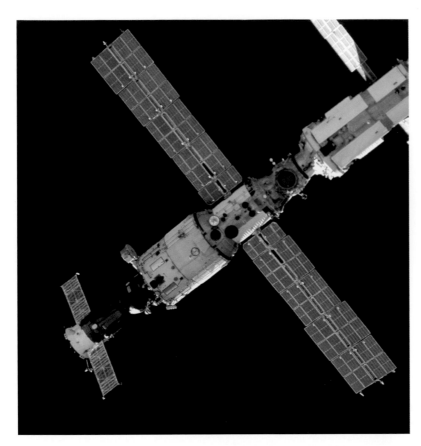

ABOVE The Russian section of the ISS as seen from below. *(NASA)*

The European Space Agency provided the Data Management Systems, the brains of Zvezda, which were responsible for controlling all module functions as well as those of the other station elements added, until those functions were taken over by the US Destiny module. Constituting the first European equipment sent as part of the ISS, the DMS was put together by an industrial consortium led by Daimler-Chrysler of Bremen, Germany. ESA supplied this equipment to the Russians in return for two flight-rated docking systems that would be used later with Europe's Automated Transfer Vehicle (ATV).

STS-106/ISS 2A.2b

Atlantis
8 September 2000, 12.46pm GMT
Cdr: Terence W. Wilcott (3)
Pilot: Scott D. Altman (1)
MS1: Edward T. Lu (1)
MS2: Richard A. Mastracchio
MS3: Daniel C. Burbank
MS4: Yuri Malenchenko (Russia)
MS5: Boris Morukov (Russia)

With two of the three assembled modules built in Russia, it was only natural that the next logistics and servicing mission to the ISS should carry two cosmonauts in its seven-man crew. *Atlantis* carried a double Spacehab module and the ICC stacked with supplies, its primary objective being to change Zvezda from the launch to the flight configuration by installing six ground repressurisation inlet caps, remove the fire extinguisher launch restraint bolts, and activate the gas masks for the first permanent resident crew (known as Expedition 1). For the first time the crew was able to offload 1,300lb (590kg) of supplies from the Progress M1-3 cargo vehicle, which had been launched on 6 August. The crew removed the TORU docking unit on Zvezda and the aft docking probe from Zarya, thus facilitating movement between the two modules. Two replacement and three new batteries were installed in Zarya along with voltage converters and other electrical equipment.

In all, 5,399lb (2,449kg) of stores were moved from Shuttle to the ISS – including 780lb

of 29lb (13kg) for attitude control. Propellant is stored in four 52.8-gallon (240-litre) tanks located in this unpressurised section, holding a total of 1,896lb (860kg) of fuel and oxidiser. Expulsion of the hypergolic propellant utilises a nitrogen pressurisation system for flow management to the motors and thrusters.

The Elektron oxygen generation system became the standard unit for supplying all compartments with oxygen but frequent problems resulted in the use of a Solid Fuel Oxygen Generator, or SFOG. More commonly known as oxygen candles, they comprise a cylindrical generator with a mixture of sodium chlorate and iron powder smouldering at 1,112°F (600°C). This produces sodium chloride, iron oxide and about 6.5 man-hours of oxygen per kilogram of stored fuel. Exhaled carbon dioxide from the crew would build up to dangerous levels if not removed and Zvezda is equipped with the Vozdukh system, which uses regenerating absorbers to scrub the air and keep it clean. Water condensates are extracted from the air by the SKV system which is then processed by the Russian Condensate Water Processor, or SRV-K.